Energy Storage

Patrick H. Stakem

(c) 2020, 2022

Table of Contents

Introduction

This is a book on energy storage. Most electrical energy is AC, and is used when generated. Direct current is stored in batterys, but there is no such things as an AC battery. There are other ways to store the energy. Solar cells only produce electricity during the day, and we might need the energy at night. Hydro power system generate power when there is a sufficient amount of water flow. Nuclear plants generate power as needed. Power from geothermal wells is generally available 24x7.

Our industrial civilization is dependent on a steady flow of electricity. But demand does not often equal supply for various reasons. We need a good energy storage system. The world's primary approach is pumped storage hydroelectric. Here, excess energy is used to pump water back "uphill," to be used later in periods of higher demand. Excess electric energy can be used to produce and store electrical power as cold or hot water. The trick is in the balancing of systems, and making them responsive to energy demand.

Iceland has many active volcanos, and has a lot of thermal energy under the surface. Enough to supply the heating and electricity needs of the country. The underground water is akin to sulfuric acid, and is not used directly. Surface water is pumped down in pipes, and returned to a turbine on the surface. The heated water is then piped underground to towns for house, businesses, and street snow clearing. There are vast resources of heat underground, and that is, in effect, their energy storage.

Storage of excess electrical energy in battery's is feasible as more efficient, less expensive units are becoming available. Generally, power systems provide alternating current, while battery's are direct current, involving two converters, one on the input side, and one on the output. Another approach is to drive a motor with a large mass spinning, or lift a heavy weight. One large system drives an electric locomotive to take heavily laden freight cars up a mountain. When the power is needed, the train drifts down, using its drive motors to back-feed the electrical system. Energy can also be stored thermally in molten salt, usually a blend of sodium nitrate and potassium nitrate. A system like this, operating at a high efficiency, basically doubles the amount of generation time. The Solana Generating Station in Arizona uses a parabolic trough solar plant. These heat water to steam by sunlight, using a reflector. It's capacity is 240 megawatts. Focused solar power heats synthetic oil to 700 degrees F. This is then used to produce steam. Storage is provided by a number of molten salt storage tanks, which are good for 6 hours.

Another approach is an ocean-going barge over a 4 kilometer-depth seafloor, winching a mass between the surface and the seabed. As the mass descends, electricity is generated. The mass is later brought up from the bottom by electric motors powered by solar cells.

Author

Mr. Stakem has a Bachelor's Degree in Electrical Engineering from Carnegie Mellon University, and Masters Degrees in Applied Physics and Computer Science from the Johns Hopkins University. He spent most of his working life at the various NASA Centers, on different space missions.

During his undergraduate years he worked summers for a local electrical facility, in the substations group. The company, at the time, was replacing the three single-phase transformers with a newer three-phase transformer. Here, he learned the hands-on of coal-fired generation, voltage step-up transformers for transmission, and "reclosers" which acted like circuit breakers at high current and voltage. In a hands-on moment, his crew chief sent him up an aluminum ladder in an active substation. Not even at the top, the fluorescent tube was growing brightly. Electric fields! They exist.

Mr. Stakem taught at the Graduate Computer Science Department of Loyola University in Maryland, Johns Hopkins Universiity, and Capitol Technology Iniversity.

Mr. Stakem can be found on Facebook and LinkedIn. Comments, corrections, suggestions are appreciated.

The Generation of Electricity

Generally, electricity is used when it is generated. We run into the scenario that we need power at night for lighting, but our solar arrays are not generating. We we have to be clever, and store energy when it is available, for use when it is needed.

No storage or source will be 100% efficient, so we will loose some of our daytime power as we draw it out at night.

In the simplest case, during the day, we]use solar generated power, and charge a battery with what we don't need. At night, we operate from the battery. We size the system so the heaviest load at night will operated properly in the hours of darkness, and that the solar panels have enough power to charge the battery's on a cloudy day.

Wind turbines generate AC power, usually 60 Hz and 600 volts. A transformer s used to supply the appropriate voltage to the grid. The turbines have a minimum "cut-in" speed, where they can start providing useful power This is around 10 MPH.

Similarly, there is a certain minimum amount of brightness for a solar panel to supply useful energy.

Today, the electrical grid is alternating current (AC). That was not always case. The very early market and technology was direct current, the favorite of inventor Thomas Edison. He was the expert, and defined the industry and the components. He got the first successful light bulb lit up in 1879. Now he could sell these to businesses and households, and then sell them the electricity to make them work. In 1880, he formed a company, the Edison Illuminating Company. His first electric Utility was the Pearl Street Station in New York City. He supplied 110 volts of direct current to some 59 customers in Manhattan. In 1882, he opened a steam-power generating station in London.

If you didn't like DC, invent your own and capitalize it. Well, a proponent of alternating current and one-time Edison employee, Nikoli Tesla did just that.

Tesla was a Serbian American who contributed greatly to the development and deployment of AC power. He became a naturalized citizen of the U. S. in 1884. He worked for Edison in New York City briefly. He later worked on wireless transmission of power, and communications.

Other early inventors such as George Westinghouse were also convinced that AC was the way of the future. In what has been called the Current Wars, both sides pointed out how dangerous and inefficient the other side's system was. Tesla's design for an poly-phase AC motor was licensed by Westinghouse. Edison's DC system only served customers within 1 mile of the generating plant. Despite the drawbacks, Edison's marketing brought lighting, refrigerators, washing machines and such to small community's across America (including the Author's home town, Cumberland, Maryland).

In what came to be called the War of the Currents, the established DC systems faced off with the upstart AC systems. Transformers were built by Westinghouse, to boost the voltage for transmission, and lower it at the customer's premises for safe use. Thinner wires were needed (although better insulation) for AC. In 1887, Westinghouse had 68 AC stations, to Edison's 121 DC stations. By 1892, the war of the currents came to an end.

Edison was bought out, and a new company, General Electric, was formed.

But, alternating current was ultimately the right answer, and dc disappeared except for some trolley systems. Locomotives used dc until switching systems could handle the high currents involved. All modern locomotives use AC power, and some are hybrid systems with a battery and inverter.

Why did ac win? Early systems were dc, because it was easier. The problem was, there is no way to distribute it widely. You might have a coal fired system for a small town, but each town needed one. It's also hard to switch large currents – you tend to melt the switch. For AC, you switch when the current goes to zero, momentarily. Out of high school, my father worked for Westinghouse, in the design of 'mercury arc" rectifiers. These could quench the big arc when switching. They were used in the conversion of ac to dc.

The Uses of Electricity

Electricity runs our world. All of our appliances (except for the gas dryer and stove) rely on electricity. We see our dependence on electricity when the power goes off. Electricity is generated in many ways, a lot of which are cheap, but cause pollution concerns. This would include coal burning plants, and nuclear plants. More recently, electrical generation by wind turbines and solar panels have come on line. Each has its pro's and con's. To make this all work efficiently, we need to be able to store energy or potential energy, to better address usage and availability.

Transmission of Electricity

Rarely do you have or even want a generating plant next to your home or businesses. Particularly nuclear (accidents do happen) or coal burning. This means we need an infrastructure of transmission equipment. As we will see later, the big problem is, the voltage needs to be a high level (100,000's of volts), to keep the transmission losses low. Power is voltage times current. With high voltage, we have a correspondingly lower current. Resistive losses in the wires is related to resistance squared. The early system and infrastructure of Thomas Edison favored direct current generation and transmission. Problem is, you generate the voltage you want, but it can only go a short distance before it dissipates as heat. As much as the early dc power systems enabled things like lighting not involving gas, or washers and dryers, you had to be within a few miles of the generating plant. You can use a gas-refrigerator, my Aunt had one, but it still needs to be plugged into the wall outlet, to run the compressor.

Electrical power transmission is inefficient for low-voltage, high current systems., For DC, you can't easily transform the power into higher or lower voltage, it has to be generated at those points. In physics, Joule's first law says the heat generated by an electrical conductor is proportional to the product of the resistance times the current squared. Thus, we want to keep the current low, and to achieve the same power, we make the voltage high. This is difficult to do with direct current. That's how it's done today with AC – the voltage is increased for transmission, then reduced at the user side.

With AC, we can readily take what we generate, boost the voltage, lower the current, and reduce the voltage back to a standard (110 volts for the US) for safe usage in user devices. The first transformer was in use in 1881.

Storage of Electricity

Technically we do not need to store electricity, but we are increasing the potential energy of air or water, or a big box of rocks, by using excess electricity at the time. We can turn this potential energy back into electricity easily, but the process is not 100% efficient. If electricity is not used when generated, it has to be stored. If a generation system is prone to the quirks of nature (as is solar and wind), then energy has to be stored when it is available, and released when it is needed. Large manufacturing plants might adjust their production schedules to take advantage of lower cost power.

Our power system, or grid, consists of sources of power, a distribution network, and users of power. If there is no storage, then electricity is used as it is generated, and the system must respond to varying loads. By 1836, the telegraph system used battery's. New York City was outfitted with a direct current infrastructure by Thomas Edison. There were storage battery's used to level the demand and supply. Pumped storage hydro was used by Connecticut Light & Power in 1929, a 31 megawatt pumped storage system. Power was also stored in flywheels. Today, pumped hydro represents the majority of power storage around the globe.

Tesla's alternating current systems proved much better at transmission, and replaced the Edison DC systems after a while. Some dc systems survived for trains and trolleys.

Direct current can be stored in battery's, but AC has to be stored in potential energy systems such as pumping water back uphill to the hydro dam. There is no free lunch. Each method of generation and storage is slightly inefficient, and some power is lost. We can convert ac to dc or vice versa with a motor-generator set, where, for example, a DC motor would drive an AC alternator. AC can be turned into DC by the rectification process. In the early trolley era, the motors were dc, and commercial AC power was used, rectified at the trolley facility. Most trolley systems today use a new generation of efficient AC motors. Locomotives, for example, use onboard diesel engines to turn alternators which supply motors at each axle. A feature called "dynamic braking" is used, when going downhill, to control speed. Essentially, the motors are driven in reverse, and generate power which is dissipated in large banks of resistor grids on the roof. This won't stop a train completely, but it does reduce brake wear.. Make a note of this, we will run into it again later.

Not before the mid 1980's did the power company's use energy storage. At that time, there were coal-fired and some nuclear plants, and expensive gas-fired peak power plants. Coal was the cheapest option, so excess power was generated and stored in a variety of methods. The coal plants did not need to be designed for the peak demand, but only the average. Pumped hydro was the preferred solution for storage.

Storage can also be implement at the customer site, where costs are controlled by purchasing and storing power when it is cheapest.

Solar systems evolved into large generators of power, as the cost of production of the cells decreased. Large "solar farms" sprung up, some with tracking arrays that followed the sun to maximize power. Wind farms also sprung up as the economics evolved, and became another non-fossil fuel, non-nuclear option. Unfortunately, solar arrays only work when the sun is shining, and wind turbines, only when the wind is blowing. Clearly, a storage system for energy is needed.

The electricity that is used for a myriad of purposes is generated as it is needed. There is not much "slack" in the generating infrastructure. Nuclear plants can adjust the output by varying the control rods in the reactor, which control the heating of water for the generators. Similarly, gas-fired plants can throttle back on the amount of gas entering the burner. Most coal-fired plants do not use raw coal, but powered coal that is blown into the combustion chamber. The control is called "load following."

Storage system can respond to varying loads much better than the generating systems.

Another issue is the sudden loss of a electrical generator or a transmission line. Today's interconnected electrical grid is fairly agile at making changes, but you might see the lights blink. Storage can smooth this out.

Sometimes, an energy storage facility at a customer's site can prevent the need for new transmission infrastructure. The customer can download a constant level of power, part of which is stored, so the generation system can operate more smoothly.

Capacitors store energy in a electric field. There is a short distance, technically, between a capacitor and an battery. An inductor stores energy in a magnetic field, as long as current flows through it. Capacitors all "leak" to a certain extent, and lose their charge over time. UltraCapacitors are very high capacity. There is a thin line between an ultra-capacitor, and a battery.

Pumped Hydro Storage Systems

This is a mature technology, in use for many decades. It uses off- peak power to essentially pump water back uphill, increasing its potential energy. As with all thermodynamic processes, it is not 100% efficient. The earliest pumped hydro facility in the United States was built in the late 1920's. The Tennessee Valley Authority (TVA) was an early adopter of this approach. There are estimated to be 40 sites in the United States, with 130 worldwide. The current practical maximum size is 4,000 megawatts, with an efficiency rating of 75-85%.

Compressed Air Storage Systems

In a compressed air storage system, a compressor is used

to pump air into a tank or underground cavern. The process is reversible, by allowing the air to turn a turbine generator. There are two operating systems of this design currently, one in Europe, and one in the U.S. Above ground storage is feasible, if there is not a convenient cave system.

The Huntorf compressed air in-ground storage facility in Northern Germany went operational in 1978. There are two separate caverns, with a total volume of over 3,00,000 cubit meters. They use a maximum pressure of 100 bar.

Battery Storage

With battery storage, you need an efficient battery chemistry, that has minimal loss on both charging ad discharging. The old technology's, lead-acid and nickle-cadmium have been superseded by Lithium-ion and other chemistry's. Batterys are only good fro a finite number of charge/discharge cycle's.

A new comer is the Sodium-sulfur battery. These are produced in commercial quantities. They do run hot, around 350 degrees c, during charge and discharge. Metallic sodium, used in the battery's, is hazardous, and combustible in the presence of water. They are rated for 4,500 charge-discharge cycles. They are used in some road vehicles as well.

Sodium-nickel-chloride battery's are in service. During charging, they produce molten sodium. Temperatures can reach 350 degrees C.

Another technology is Vanadium Redox (reduction and oxidation). The vanadium is in an aqueous solution of sulfuric acid, like a lead-acid battery. The operating life is not yet known, but estimated at 10 years. They use pumps to circulate the electrolyte.

More advanced battery technologies are in development and test, including the Iron-chromium battery, zinc-bromine, zinc-air, lead acid-carbon, and others.

Lithium-ion batteries are a recent technology, only having seen widespread use in the last couple of years. They have found a home in hybrid and all-electric vehicles. Tesla, the big producer of electric cars and a truck, has a stationary unit called the Power Wall, for home application. It is supposed to handle your electrical needs for 7 days of power outage. It is easily paired with a solar systems.

Lithium Ion battery's of this size are quoted as having a 15 year life, and are currently priced around $6,500.

The largest battery energy storage at the moment is the Hornesdale Power Reserve in South Australia. It uses a Tesla Powerpack system to hold 100 megawatts of power for peak demand. The Powerpack provides more than 2oo Kilowatt-hours of storage capability

The Tesla Powerwall is a stationary energy storage device, using the same Lithium-ion battery's developed for the cars. It was announced in 2015, with a 10 kilowatt-hour capacity. It uses liquid cooling to keep the unit within safe and efficient temperature levels. The Powerwall has a 92.5% efficiency, and is good for 5,000 cycles.

Flywheel Storage

In flywheel storage, we have a large spun flywheel, with a large amount of angular momentum. It stores energy by increasing its rotation rate with a motor, and provides energy when the rotating wheel drive a generator. A major issue for large flywheel system is containment of heavy, fast moving pieces should the flywheel fail mechanically. They are typically enclosed in a vacuum to reduce air drag. They produce dc current, which needs to be changed into ac by an inverter, for use on the grid. Efficiency's are around 80%, and they have power density's of almost 10 times that of battery's

There is an existing system in New England, The Beacon New York Flywheel energy storage plant, storing a megawatt of power. 200 flywheels are situated on 5 acres. Each unit has a capacity of 100 kw. Mass production is driving down costs for such systems, and the use of composite materials make them stronger.

Ice Storage

This is sometimes used as an adjunct to air conditioning, producing ice when the electrical rates are lowest, and cooling a building with the ice. There is one in California, Redding Electric Utility's, that is reaching peak load reduction by 6 megawatts.

Superconducting magnetic energy storage

This approach uses a superconducting coil, which has been cooled below its superconducting critical point. The energy is stored in a magnetic field. Its a basically loss-less system, as long as the temperature is kept low enough. Energy is used to keep the device below its superconducting threshold.

The power in the coil is direct current. This causes a small energy loss on the input and output sides, as ac is converted to dc, and then back. The overall efficiency of the system can be as high as 95%.

A test facility in Wisconsin has a storage capacity of 20 megawatts.In different configurations, the temperatures range from several degrees to several tens of degrees absolute.

New concepts

A variety of new efficient energy storage systems are in design or test. These involve new technologys, and are evaluated for their relative costs and benefits.

Some of these include:

1. improvements on lead-acid batteries, including deep discharge.

2. supercapacitors – 10-100 times the energy storage of a regular capacitor, and more charge-discharge cycles.

3. liquid air storage – storing energy in a

cryogenic fluid.

4. advanced flywheels - using high strength composite materials.

5. new battery chemistry's and materials.

Alternating Current controls the market, but now we have many different ways to generate power, and to store it until needed. If you want to go off the grid, with solar or wind power, there is a storage solutions available.

Afterword

The ability to store large amounts of energy when it is available, and use it when needed, affects the economics of many new generation systems. Expect to see wider uses, and new techniques that expand the energy horizons.

Glossary of Terms

AC "battery" dc battery and an inverter, packaged together.

Active power - "real power" in watts. Direction of energy flow does not reverse.

Ampere – a unit of current flow.

Apparent Power – product of the rms value of voltage and current.

APU – auxiliary power unit

ASIN – Amazon Standard Inventory Number.

BESS – Battery Energy Storage System.

Bar – unit of pressure, 100 Kpa, atmosphere pressure at 111 meters and 15 degrees C.

Black start – ability of a facility to start without external power.

BTES – Borehole thermal energy storage.

BTU – British Thermal Unit

CAES – Compressed air energy storage.

Capacitor – stores energy in an electric field.

CES – Community Energy Storage.

CSP – concentrated solar power.

DC – direct current; flows in one direction only.

Deep cycle battery – can be regularly discharged to close to capacity, dod – depth of discharge.

DOE – (U. S.) Department of Energy.

EPRI – Electric Power Research Institute.

EsaaS – energy storage as a service.

FERC – (U. S.) Federal Energy Regulatory Commission.

FES – Flywheel energy storage.

Flow battery – having two chemical compounds separated by a membrane.

Fuel cell – spent fuel is extracted, and new fuel is added.

GESDB – (U. S. DOE) Global Energy Storage Database.

GST – grid storage technology.

Heat Pump – moves energy from a source to a thermal reservoir.

IESDB – (U.S. DOE) International Energy Storage Database.

Inductor – stores energy in a magnetic field, as current flows through it.

Inverter – converts dc to AC.

Joule – Unit of energy – one Newton per meter.

Kinetic Energy – energy of motion.

Kw – kilowatt.

kwh - kilowatt-hour

LAES – Liquid air energy storage.

LHTES – Latent heat thermal energy storage.

LMS – load management system.

MW – Mega (10^6) watt.

Non-spinning reserve– offline energy reserve.

PCM – phase change material.

PHES – Pumped hydroelectric energy storage.

Potential Energy – energy of an object because of its position, its charge, its temperature, etc.

Power factor – ratio of actual electric power in an AC circuit to the product of the rms values of the voltage and current. Difference due to reactance in the circuit.

Primary battery – single use. Not rechargeable.

Reactance – non-resistance impedance in an AC circuit. Current is out of phase with the voltage.

Reactive power – voltage and current are 90 degrees out of phase, due to the nature of the load.

RMS – root mean square.

Secondary battery – rechargeable.

SMES – superconducting magnetic energy storage.

SoC – state of charge.

Spinning reserve – generation capacity that is online, but not connected.

STES – Seasonal thermal energy storage.

TES – Thermal energy storage.

Ultracapacitor – high capacity capacitor.

UPS – uninterruptible power supply.

VAR – volts-ampere reactive.

Volt – a unit of electromotive force.

Watt – a unit of power in the Metric System. Units of one Joule per second.

References

Asian Development Bank, *Handbook on Battery Energy Storage System*, 2018, ISBN-978-9292614706.

Barnes. Frank S., Levine, Jonah G. *Large Energy Storage Systems Handbook,* 2011, CRC Press, avail: Google Play

Beard, Kirby W. *Linden's Handbook of Batteries*, Fifth Edition, 2019, McGraw Hill, ISBN-978-1260115925.

Blume, Steven W. *Electric Power System Basics for the Nonelectrical Professional,* 2016, ISBN-978-1119180197.

Breeze, Paul *Power System Energy Storage Technologies* (The Power Generation), 1st Edition, Academic Press, 2018, ISBN-978-0128129029.

Cawthorne, Nigel, *Tesla vs Edison: The Life-Long Feud that Electrified the World,* 2016, ISBN-978-0785833789.

Cline, Adam *The Current War: A Battle Story Between Two Electrical Titans, Thomas Edison And George Westinghouse*, 2017, ASIN-B076B576QD.

Crawley, Gerald M. *Energy Storage* (World Scientific Series in Current Energy Issues), 2017, ISBN-978-9813208957.

Dawes, Chester L. *A Course in Electrical Engineering,* Vol. 1, *Direct Currents;* McGraw Hill, 1927, Maple Press; reprinted 2015, ISBN-978-1296722043. (This was my Father's text when he attended Carnegie Institute of Technology in 1928, These is a Volume II, *Alternating Current.*)

Huggins, Robert *Energy Storage: Fundamentals, Materials, and Applications*, 2016, ISBN-978-3319212388.

Jones, Lawrence E. *Renewable Energy Integration: Practical Management of Variability, Uncertainty, and Flexibility in Power Grids,* 2017, ISBN-978-0128095928.

Jones, Jill *Empires of Light: Edison, Tesla, Westinghouse, and the Race to Electrify the World,* 2004, ISBN-978-0375758843.

Leupp, Francis Ellington, *George Westinghouse: His Life and Achievements (1918),* 2018, ASIN: B078Z87T62.

Rufer, Alfred *Energy Storage: Systems and Components,* 2017, ISBN-978-1138082625.

Slack, Edgar P. *Elementary Electricity,* 1943, McGraw-Hill, ASIN-B003TK5CRE.

Smit, Berend et al, *Introduction to Carbon Capture and Sequestration,* 2014, ISBN-978-1783263288.

Sparber, Andrew G. *The Electrification of Allegany County, MD,* 2018, ISBN- 978-1978170582.

Sterner, Michael; Stadler, Ingo *Handbook of Energy Storage: Demand, Technologies, Integration,* 2019, ISBN-978-3662555033.

Stross, Randell E. *The Wizard of Menlo Park: How Thomas Alva Edison Invented the Modern World,* 2007, ASIN-B000OI0G9S.

Tesla, Nikola, *My Inventions: The Autobiology of Nikola Tesla,* 2018, ISBN-978-1684222063.

Wu, Fu-Bao, et al, *Grid-Scale Energy Storage Systems and Applications,* 1st Edition, Academic Press, 2019, ISBN-978-0128152928.

Zini, Gabriele *Green Electrical Energy Storage: Science and Finance for Total Fossil Fuel Substitution,* 1st Edition, 2016, McGraw Hill, ISBN-978-1259642838.

Resources

Sandia National Labratories, *DOE/EPRI Electricity Storage Handbook,* avail: https://www.sandia.gov/ess-ssl/handbook.php

https://www.sandia.gov/ess-ssl/global-energy-storage-database/

http://www.electricitystorage.org/about/about_esa

http://www.storagealliance.org

http://www.sandia.gov/eesat/

http://www.ease-storage.eu/

http://energytoolbase.com

http://www.conedsolutions.com

http://www.energystorage.com

http://www.energytoolbase.com

http:..www,phasechange.com

http://wnwergy-storage.news

www.energy.gov

http://www.edf.org/energystorage/emissions
Wikipedia, various..

If you enjoyed this book, you might also be interested in some of these.

Stakem, Patrick H. *16-bit Microprocessors, History and Architecture*, 2013 PRRB Publishing, ISBN-1520210922.

Stakem, Patrick H. *4- and 8-bit Microprocessors, Architecture and History*, 2013, PRRB Publishing, ISBN-152021572X,

Stakem, Patrick H. *Apollo's Computers,* 2014, PRRB Publishing, ISBN-1520215800.

Stakem, Patrick H. *The Architecture and Applications of the ARM Microprocessors,* 2013, PRRB Publishing, ISBN-1520215843.

Stakem, Patrick H. *Earth Rovers: for Exploration and Environmental Monitoring,* 2014, PRRB Publishing, ISBN-152021586X.

Stakem, Patrick H. *Embedded Computer Systems, Volume 1, Introduction and Architecture*, 2013, PRRB Publishing, ISBN-1520215959.

Stakem, Patrick H. *The History of Spacecraft Computers from the V-2 to the Space Station*, 2013, PRRB Publishing, ISBN-1520216181.

Stakem, Patrick H. *Floating Point Computation*, 2013, PRRB Publishing, ISBN-152021619X.

Stakem, Patrick H. *Architecture of Massively Parallel Microprocessor Systems*, 2011, PRRB Publishing, ISBN-1520250061.

Stakem, Patrick H. *Multicore Computer Architecture,* 2014, PRRB Publishing, ISBN-1520241372.

Stakem, Patrick H. *Personal Robots*, 2014, PRRB Publishing, ISBN-1520216254.

Stakem, Patrick H. *RISC Microprocessors, History and Overview,* 2013, PRRB Publishing, ISBN-1520216289.

Stakem, Patrick H. *Robots and Telerobots in Space Applications*, 2011, PRRB Publishing, ISBN-1520210361.

Stakem, Patrick H. *The Saturn Rocket and the Pegasus Missions, 1965,* 2013, PRRB Publishing, ISBN-1520209916.

Stakem, Patrick H. *Visiting the NASA Centers, and Locations of Historic Rockets & Spacecraft,* 2017, PRRB Publishing, ISBN-1549651205.

Stakem, Patrick H. *Microprocessors in Space*, 2011, PRRB Publishing, ISBN-1520216343.

Stakem, Patrick H. Computer *Virtualization and the Cloud*, 2013, PRRB Publishing, ISBN-152021636X.

Stakem, Patrick H. *What's the Worst That Could Happen? Bad Assumptions, Ignorance, Failures and Screw-ups in Engineering Projects, 2014,* PRRB Publishing, ISBN-1520207166.

Stakem, Patrick H. *Computer Architecture & Programming of the Intel x86 Family, 2013,* PRRB Publishing, ISBN-1520263724.

Stakem, Patrick H. *The Hardware and Software Architecture of the Transputer,* 2011,PRRB Publishing, ISBN-152020681X.

Stakem, Patrick H. *Mainframes, Computing on Big Iron,* 2015, PRRB Publishing, ISBN- 1520216459.

Stakem, Patrick H. *Spacecraft Control Centers,* 2015, PRRB Publishing, ISBN-1520200617.

Stakem, Patrick H. *Embedded in Space,* 2015, PRRB Publishing, ISBN-1520215916.

Stakem, Patrick H. *A Practitioner's Guide to RISC Microprocessor Architecture,* Wiley-Interscience, 1996, ISBN-0471130184.

Stakem, Patrick H. *Cubesat Engineering,* PRRB Publishing, 2017, ISBN-1520754019.

Stakem, Patrick H. *Cubesat Operations*, PRRB Publishing, 2017, ISBN-152076717X.

Stakem, Patrick H. *Interplanetary Cubesats*, PRRB Publishing, 2017, ISBN-1520766173 .

Stakem, Patrick H. Cubesat Constellations, Clusters, and Swarms, Stakem, PRRB Publishing, 2017, ISBN-1520767544.

Stakem, Patrick H. *Graphics Processing Units, an overview,* 2017, PRRB Publishing, ISBN-1520879695.

Stakem, Patrick H. *Intel Embedded and the Arduino-101, 2017,* PRRB Publishing, ISBN-1520879296.

Stakem, Patrick H. *Orbital Debris, the problem and the mitigation,* 2018, PRRB Publishing, ISBN-*1980466483*.

Stakem, Patrick H. *Manufacturing in Space,* 2018, PRRB Publishing, ISBN-1977076041.

Stakem, Patrick H. *NASA's Ships and Planes,* 2018, PRRB Publishing, ISBN-1977076823.

Stakem, Patrick H. *Space Tourism,* 2018, PRRB Publishing, ISBN-1977073506.

Stakem, Patrick H. *STEM – Data Storage and Communications,* 2018, PRRB Publishing, ISBN-

1977073115.

Stakem, Patrick H.   *In-Space Robotic Repair and Servicing*, 2018, PRRB Publishing, ISBN-1980478236.

Stakem, Patrick H. *Introducing Weather in the pre-K to 12 Curricula, A Resource Guide for Educators*, 2017, PRRB Publishing, ISBN-1980638241.

Stakem, Patrick H. *Introducing Astronomy in the pre-K to 12 Curricula, A Resource Guide for Educators*, 2017, PRRB Publishing, ISBN-198104065X.
Also available in a Brazilian Portuguese edition, ISBN-1983106127.

Stakem, Patrick H. *Deep Space Gateways, the Moon and Beyond*, 2017, PRRB Publishing, ISBN-1973465701.

Stakem, Patrick H. *Exploration of the Gas Giants, Space Missions to Jupiter, Saturn, Uranus, and Neptune*, PRRB Publishing, 2018, ISBN-9781717814500.

Stakem, Patrick H. *Crewed Spacecraft*, 2017, PRRB Publishing, ISBN-1549992406.

Stakem, Patrick H. *Rocketplanes to Space*, 2017, PRRB Publishing, ISBN-1549992589.

Stakem, Patrick H. *Crewed Space Stations,* 2017, PRRB Publishing, ISBN-1549992228.

Stakem, Patrick H. *Enviro-bots for STEM: Using Robotics in the pre-K to 12 Curricula, A Resource Guide for Educators,* 2017, PRRB Publishing, ISBN-1549656619.

Stakem, Patrick H. *STEM-Sat, Using Cubesats in the pre-K to 12 Curricula, A Resource Guide for Educators*, 2017, ISBN-1549656376.

Stakem, Patrick H. *Lunar Orbital Platform-Gateway*, 2018, PRRB Publishing, ISBN-1980498628.

Stakem, Patrick H. *Embedded GPU's*, 2018, PRRB Publishing, ISBN- 1980476497.

Stakem, Patrick H. *Mobile Cloud Robotics*, 2018, PRRB Publishing, ISBN- 1980488088.

Stakem, Patrick H. *Extreme Environment Embedded Systems,* 2017, PRRB Publishing, ISBN-1520215967.

Stakem, Patrick H. *What's the Worst, Volume-2*, 2018, ISBN-1981005579.

Stakem, Patrick H., *Spaceports*, 2018, ISBN-1981022287.

Stakem, Patrick H., *Space Launch Vehicles*, 2018, ISBN-1983071773.

Stakem, Patrick H. *Mars*, 2018, ISBN-1983116902.

Stakem, Patrick H. *X-86, 40th Anniversary ed*, 2018, ISBN-1983189405.

Stakem, Patrick H. *Lunar Orbital Platform-Gateway*, 2018, PRRB Publishing, ISBN-1980498628.

Stakem, Patrick H. *Space Weather*, 2018, ISBN-1723904023.

Stakem, Patrick H. *STEM-Engineering Process*, 2017, ISBN-1983196517.

Stakem, Patrick H. *Space Telescopes*, 2018, PRRB Publishing, ISBN-1728728568.

Stakem, Patrick H. *Exoplanets*, 2018, PRRB Publishing, ISBN-9781731385055.

Stakem, Patrick H. *Planetary Defense*, 2018, PRRB Publishing, ISBN-9781731001207.

Patrick H. Stakem *Exploration of the Asteroid Belt*, 2018, PRRB Publishing, ISBN-1731049846.

Patrick H. Stakem *Terraforming*, 2018, PRRB Publishing, ISBN-1790308100.

Patrick H. Stakem, *Martian Railroad*, 2019, PRRB Publishing, ISBN-1794488243.

Patrick H. Stakem, *Exoplanets,* 2019, PRRB Publishing, ISBN-1731385056.

Patrick H. Stakem, *Exploiting the Moon,* 2019, PRRB Publishing, ISBN-1091057850.

Patrick H. Stakem, *RISC-V, an Open Source Solution for Space Flight Computers,* 2019, PRRB Publishing, ISBN-1796434388.

Patrick H. Stakem, *Arm in Space*, 2019, PRRB Publishing, ISBN-9781099789137.

Patrick H. Stakem, *Extraterrestrial Life*, 2019, PRRB Publishing, ISBN-978-1072072188.

Patrick H. Stakem, *Space Command*, 2019, PRRB Publishing, ISBN-978-1693005398.

CubeRovers, A Synergy of Technologys, 2020, PRRB Publishing, ISBN-979-8651773138.

Robotic Exploration of the Icy moons of the Gas Giants. 2020, PRRB Publishing, ISBN- 979-8621431006

Hacking Cubesats, 2020, PRRB Publishing, ISBN-979-8623458964.

History & Future of Cubesats, PRRB Publishing, ISBN-979-8649179386.

Hacking Cubesats, Cybersecurity in Space, 2020, PRRB Publishing, ISBN-979-8623458964.

Powerships, Powerbarges, Floating Wind Farms: electricity when and where you need it, 2021, PRRB Publishing, ISBN-979-8716199477.

Hospital Ships, Trains, and Aircraft, 2020, PRRB Publishing, ISBN-979-8642944349.

2020/2021 Releases

CubeRovers, a Synergy of Technologys, 2020, ISBN-979-8651773138

Exploration of Lunar & Martian Lava Tubes by Cube-X, ISBN-979-8621435325.

Robotic Exploration of the Icy moons of the Gas Giants, ISBN- 979-8621431006.

History & Future of Cubesats, ISBN-978-1986536356.

Robotic Exploration of the Icy Moons of the Ice Giants, by Swarms of Cubesats, ISBN-979-8621431006.

Swarm Robotics, ISBN-979-8534505948.

Introduction to Electric Power Systems, ISBN-979-8519208727.

Centros de Control: Operaciones en Satélites del Estándar CubeSat (Spanish Edition), 2021, ISBN-979-8510113068.

Exploration of Venus, 2022, ISBN-979-8484416110.

Patrick H. Stakem, *The Search for Extraterrestial Life,* 2019, PRRB Publishing, ISBN-1072072181.

The Artemis Missions, Return to the Moon, and on to Mars, 2021, ISBN-979-8490532361.

James Webb Space Telescope. A New Era in Astronomy, 2021, ISBN-979-8773857969.